Toxic Waste

Peggy J. Parks

KIDHAVEN PRESS
A part of Gale, Cengage Learning

Detroit • New York • San Francisco • New Haven, Conn • Waterville, Maine • London

© 2006 Gale, Cengage Learning

For more information, contact
KidHaven Press
27500 Drake Rd.
Farmington Hills, MI 48331-3535
Or you can visit our Internet site at gale.cengage.com

ALL RIGHTS RESERVED.
No part of this work covered by the copyright hereon may be reproduced or used in any form or by any means—graphic, electronic, or mechanical, including photocopying, recording, taping, Web distribution or information storage retrieval systems—without the written permission of the publisher.

Every effort has been made to trace the owners of copyrighted material.

LIBRARY OF CONGRESS CATALOGING-IN-PUBLICATION DATA

Parks, Peggy J., 1951–
   Toxic waste / by Peggy J. Parks.
      p. cm. — (Our environment)
   Includes bibliographical references and index.
   Contents: Hazardous ooze—How serious is the problem?—Toxic waste disasters—Is there an answer?
   ISBN 0-7377-1823-4 (hardcover : alk. paper) 1. Hazardous wastes—Juvenile literature. 2. Hazardous wastes—Environmental aspects—Juvenile literature. I. Title. II. Series.
   TD1030.5.P37 2006
   363.72'87—dc22
                                                         2005026848

Printed in the United States of America
4 5 6 7 12 11 10 09 08

contents

Chapter 1
Hazardous Ooze 4

Chapter 2
How Serious Is the Problem? 13

Chapter 3
Toxic Waste Disasters 22

Chapter 4
Is There an Answer? 30

Notes 39

Glossary 41

For Further Exploration 43

Index 45

Picture Credits 48

About the Author 48

chapter one

Hazardous Ooze

Every day of the year, staggering amounts of waste products are generated throughout the world. This includes everything from mountains of office paper to tons of restaurant garbage, and from junk automobiles to worn-out tires. In the United States alone, billions of tons of waste are created annually. It is such an enormous amount that if all the waste could be piled into garbage trucks, the line of trucks could stretch halfway to the moon! Apart from clogging the world's overflowing landfills, most waste is not necessarily harmful—but that is certainly not true of all of it. Many types of waste are poisonous to the environment, humans, and wildlife. This is known as **toxic waste**.

The Role of Industry

Although toxic waste comes from numerous sources, the largest amount is generated by industries. Steel mills, metal plating plants, furniture factories, and petroleum refineries produce various forms of toxic waste. Other contributors include plastic manufacturers, companies that make leather products, and producers of chemicals. Basically, almost every type of industry that exists generates some form of toxic waste.

A massive mound of garbage offers a reminder of the huge amount of waste humans generate.

6   Toxic Waste

Paper mills, which use many different chemicals and acids in the paper-making process, are some of the world's largest producers of toxic waste. Many of these mills use chlorine to bleach wood pulp before it is made into paper. In the process, chemical by-products called **dioxins** are created. Dioxins are some of the deadliest substances in existence. They can enter the atmosphere through the mills' huge smokestacks, or be washed into lakes, streams, and wetlands through wastewater. Besides paper mills, other industrial producers of dioxins include city waste operations, cement manufacturers, and coal-fired power plants.

Paper mills (below) and pulp mills (inset) produce many different kinds of toxic waste.

Another group of extremely toxic substances is **heavy metals.** Mercury, cadmium, lead, chromium, and arsenic are all examples of these metals. They are generated by many industrial sources, such as factories that make motor vehicle parts, metal furniture, or photographic equipment. Extremely high concentrations of lead are often found in the **sludge**, or muddy waste products, of battery manufacturing plants. Power plants emit tons of mercury into the atmosphere when they burn coal to make electricity. Once the mercury is in the air, the wind can carry it over great distances. According to the United Nations Environment Programme (UNEP), most of the mercury found in Florida's Everglades National Park comes from Europe and Africa, thousands of miles away.

Toxic Mining

One of the largest producers of toxic metals in waste products is the mining industry. When minerals such as gold, platinum, silver, or copper are removed from the earth, mining companies often use the "open-pit" method. The process starts with powerful explosives that blast away rocks and soil. Then the rock rubble is heaped into enormous piles. Contained in the rubble are significant amounts of mercury, arsenic, and lead that exist naturally within rock. Rain and snow fall on the rock piles, washing the toxic substances into the soil and surface water. The toxins can also seep

Rubble created by copper mines, such as this one in Canada, contains toxic substances that can end up in the soil and water.

deep into the ground to pollute water supplies known as **groundwater**.

Another way that mining operations generate toxic waste is through a process known as **heap leaching**. This technique is used to separate minerals from ore, the rock that encases them. Using heavy equipment, mine workers pile crushed ore into mounds so gigantic they could cover an entire football field. Then they spray the ore with a poisonous chemical called cyanide. The cyanide seeps through the ore and bonds with the gold or other mineral. The liquid solution is collected, and other chemicals, such as mercury, are added to filter out the minerals. The cyanide mixture is then stored in artificial ponds to be reused later. These deadly toxic ponds are of great concern to environmental scientists because of the potential for leakage and major spills.

E-Waste

Cyanide, dioxins, and other toxic substances are often generated by industry during the production of new products. But toxic waste can also end up in the environment when old products are thrown out. For instance, more than ever before, people are upgrading their computers, cellular phones, televisions, and other electronic equipment. Disposal of the old products creates what is known as **e-waste**. According to Greenpeace International, as much as 55 million tons (50 million metric tons) of e-waste are generated annually. Such an enormous amount is hard to imagine, so Greenpeace offers an illustration: "Think of it like this—if the estimated amount of e-waste generated every year would be put into containers on a train it would go once around the world!"[1]

E-waste is toxic because electronic devices contain several hundred different materials, including heavy metals. For instance, lead is used in computer circuit boards, as well as in the cathode-ray tubes (CRTs) in monitors. Batteries for computers and cell phones contain high concentrations of lead, mercury, and cadmium, while lighting devices for flat-screen monitors contain mercury.

10 Toxic Waste

E-waste is now the fastest-growing type of solid waste, and it is expected to increase even more in the future. This is creating a serious problem in landfills, where the discarded electronics usually end up. After long periods of time, the equipment decomposes, or breaks down. The toxic metals inside them can leak into the soil and eventually seep into groundwater.

Common Toxins

Many other toxic waste products end up in landfills as well. For instance, people often toss old batteries from cars, flashlights, or toys into the trash. When the batteries pile up in landfills, lead and acids can leak into the soil. In addition, potentially

Members of the environmental group Greenpeace protest toxic electronic waste in Guadalajara, Mexico.

toxic household products are often disposed of improperly. For instance, old paint, pesticides, prescription drugs, and cleaning solutions are often dumped down the drain. The toxic substances end up in sewers and eventually wash into bodies of water such as streams, lakes, and rivers.

One of the most toxic common-waste products is used motor oil. When people change the oil in their vehicles, many of them dispose of the used oil by dumping it on the ground or pouring it into drains or sewers. Every year an estimated 400 million gallons (1.5 billion liters) of used motor oil is disposed of in this way. That is about twenty times as much as the oil spilled in the worst supertanker accidents. Also, some ships deliberately dump used oil into the ocean. According to the National Academy of Sciences, millions of gallons of oil sludge are illegally dumped into the ocean each year. Because oil contains high levels of lead and compounds known as **hydrocarbons**, it is highly toxic to the environment. Just 1 gallon (3.8 liters) of used motor oil can pollute a million gallons (3.8 million liters) of drinking water!

A Growing Problem

Used motor oil, e-waste, and chemical by-products from industry are only a few examples of toxic waste. There are many others as well. Weapons manufacturers and power plants produce **radioactive waste**, or waste that is generated during the production of nuclear energy. Toxic waste from hospitals and

Nuclear power plants such as this one produce radioactive waste that must be stored in special containers (inset).

other health-care facilities includes bloody bandages, used needles and syringes, and diseased organs from humans and animals. On farms, toxic waste is created when livestock manure, fertilizers, herbicides, and pesticides wash into wetlands and streams.

All in all, there are thousands of types of toxic waste generated by many sources all over the world. As the population continues to grow, the amount of toxic waste will undoubtedly grow as well—and that means it could become an even greater problem in the future.

chapter two

How Serious Is the Problem?

For years, scientists have known about the risks of toxic waste. Although they do not always agree on how serious the problem is, research has shown that many toxic substances are harmful. For instance, mercury, lead, cadmium, and other heavy metals can damage the brain, nervous system, blood vessels, and kidneys. Cyanide can also cause brain damage, and high amounts can lead to unconsciousness or death. Aluminum has been associated with diseases such as Alzheimer's and Parkinson's. Dioxins have been linked to a number of health problems, including skin diseases, liver damage, and cancer.

14   Toxic Waste

These and other toxic substances enter the environment in many different ways. One of the most common sources is leakage, either from landfills or hazardous waste disposal sites. There is no way to know the full extent of the problem because scientists do not always know where the waste is buried, or whether it is leaking. In the state of Kentucky, for instance, a hazardous waste disposal area called the

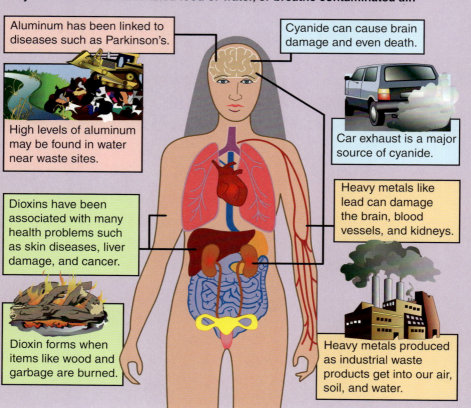

Effects of Toxic Waste on the Body

Toxic wastes can pollute air, soil, and water. People are at risk when they eat or drink contaminated food or water, or breathe contaminated air.

Aluminum has been linked to diseases such as Parkinson's.

High levels of aluminum may be found in water near waste sites.

Dioxins have been associated with many health problems such as skin diseases, liver damage, and cancer.

Dioxin forms when items like wood and garbage are burned.

Cyanide can cause brain damage and even death.

Car exhaust is a major source of cyanide.

Heavy metals like lead can damage the brain, blood vessels, and kidneys.

Heavy metals produced as industrial waste products get into our air, soil, and water.

Valley of the Drums was discovered in 1967. More than 17,000 rusting, leaking barrels filled with toxic waste were illegally scattered about the land. Another area in Kentucky known as Maxey Flats is among the largest radioactive waste sites in the world. The facility was shut down in 1977 when radioactive material was found leaking from trenches where the waste was buried. Kentucky environmental officials have said that Maxey Flats is one of the state's "biggest environmental nightmares and . . . the problems associated with this site will haunt Kentuckians for many generations to come."[2]

How Toxins Affect Living Things

No matter how toxic substances enter the environment, they can pollute soil, water, and air. And that is when they become serious hazards. Plants absorb toxins in the soil and water through their roots, stems, and leaves. Animals eat the plants, which can cause toxins to build up in their fatty tissues. Humans can get the substances in their bodies by breathing toxins in the air, drinking polluted water, or eating contaminated plants and animals.

Scientists say that people who have the highest mercury levels are those who eat large quantities of fish. Mercury tends to settle in the sediments of lakes, rivers, and streams. Microscopic plants and animals at the bottom of the **food chain** absorb the toxins. Small fish eat the tiny organisms and mercury builds up in their tissue. Bigger fish eat the smaller fish and absorb even higher concentrations of mercury. Predator fish, such as swordfish, barracuda, and tuna, are usually the most contaminated because their diet consists of smaller fish that are often full of mercury.

In some areas of the world such as the Arctic, seals and other marine mammals are a major part of people's diet, along with fish. According to UNEP, the risks of mercury contamination are much higher in those areas than other countries. In Greenland, for instance, mercury levels in fish and marine mammals are three times higher today than they were two centuries ago. As a result, more than 16 percent of the people living in northern Greenland have abnormally high levels of mercury in their blood.

Toxic Farming

Although mercury and other heavy metals are most common in fish, they can also find their way into farm crops and livestock. One source is fertilizer that has been made from toxic waste. A 2001 report by the California Public Interest Research Group (CALPIRG) states that this type of fertilizer is extremely harmful because toxic substances contaminate the land. CALPIRG's research involved testing 29 fertilizers from 12 different states. The group found that the fertilizers contained more than 20 heavy metals including arsenic, cadmium, lead, and mercury. One fertilizer from Michigan also contained dioxins.

Toxic waste is used for fertilizers in other countries as well. For instance, a carbonated-soda plant in Kerala, India, regularly provides its waste sludge to local farmers. Investigators took samples of the sludge and sent them to Great Britain for examination. The tests revealed that the waste material contained many toxic metals, including cadmium and lead. John Henry, a British expert on poisons, says the practice of using toxic waste for fertilizer should be banned immediately. He explains: "The results have devastating consequences for those living near the areas where this waste has been dumped and for the thousands who depend on

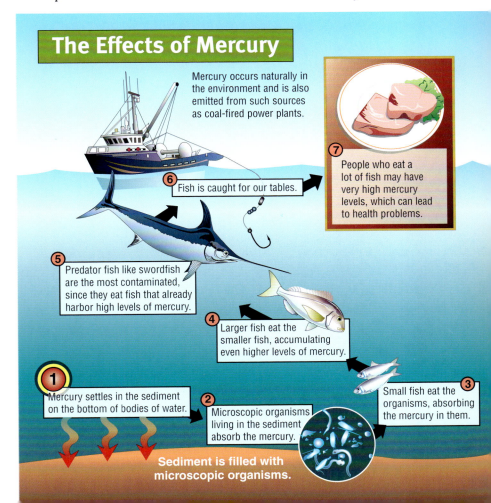

18   Toxic Waste

crops produced in these fields."[3] British scientists say that the contamination is not limited to the soil. It has spread into Kerala's water supply, causing high levels of lead in the city's drinking water.

The Tiniest Victims

Lead and other toxic substances are dangerous for people of all ages. The greatest risk, however, is to babies and children. When children are very young, their brains, lungs, and nervous systems are still developing. The organs' small size makes

Donna Robins, like some other parents in Woburn, Massachusetts, lost a child to leukemia caused by exposure to toxic waste. She hugs her other son in a 1980 photo.

How Serious Is the Problem? 19

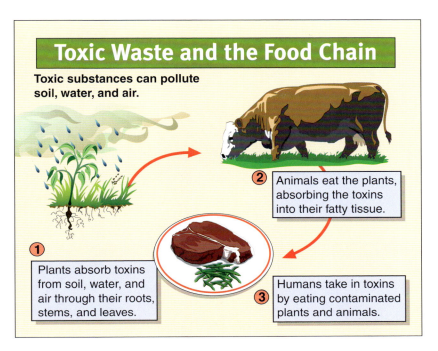

them more vulnerable to damage. Also, immune systems, which naturally fight off toxins, are not fully developed in babies and young children.

One tragic case of toxic waste causing childhood disease occurred in the town of Woburn, Massachusetts. Between 1966 and 1986, 28 Woburn children were diagnosed with leukemia, a cancer of the blood. That number was abnormally high—over four times higher than the national average for a community that size. The people of Woburn did not know that their wells had been poisoned with heavy metals, such as arsenic, lead, chromium, and mercury, as well as other toxic chemicals. Through a combination of illegal dumping, accidental spills, and leakage, several industrial operations had severely polluted the soil and groundwater. When pregnant women drank the contaminated water, they passed the toxins along to their unborn babies.

The Massachusetts Department of Public Health (DPH) extensively studied the Woburn case. In 1996, the agency concluded that there was a strong connection between the high rate of leukemia and the polluted water. The case was controversial because proving that toxic substances actually cause human disease can be challenging. However, DPH official Suzanne Condon said there was clearly a connection, as she explained: "Particularly in Woburn, how anybody could believe that what happened was by chance is ridiculous."[4]

Toxic waste from industry can harm the human fetus (inset) if the waste gets into the water supply.

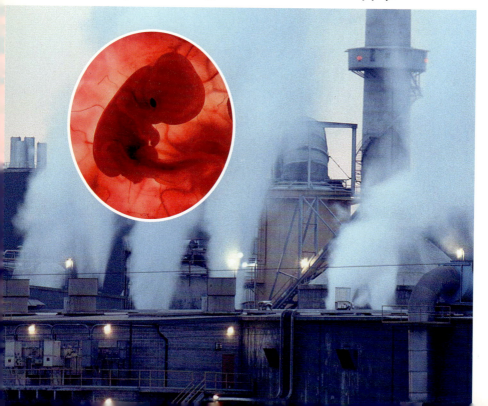

Risk to the Unborn

One reason the Woburn case was challenging was because the children were contaminated before they were born. For years scientists were convinced that fetuses were protected from toxins in the mother's blood. However, a September 2004 study by the Environmental Working Group (EWG) proved that theory to be wrong. The study confirmed that unborn babies can indeed have an alarming variety of toxic substances in their blood.

Researchers tested blood from ten newborn infants at hospitals throughout the United States. They found that the blood contained 287 different toxic substances, including mercury, dioxins, and polychlorinated biphenyls (commonly known as PCBs). The EWG explains why these findings are so disturbing: "The fact is, a child can bear a lifelong imprint of risks from the countless . . . industrial pollutants that find their way through the placenta, down the umbilical cord, and into the baby's body. The consequences—health disorders, subtle or serious—can surface not only in childhood but also in adulthood."[5]

Scientists may disagree about how serious the toxic waste problem is, but one thing cannot be disputed: Numerous toxic substances are harmful to humans. Whether the toxins come from fish, polluted water, or crops grown in contaminated soil, they pose a serious health risk for people all over the world.

chapter three

Toxic Waste Disasters

There have been many occasions when toxic waste has harmed the environment. The causes include everything from contaminated fertilizer to leakage in hazardous waste landfills or disposal tanks. In some instances, major spills or other accidents have occurred, and have caused widespread damage. Most of those were the result of human actions, but others involved natural disasters.

Waves of Waste

One of history's worst natural disasters occurred in December 2004, and it resulted in a toxic waste crisis. The Indian Ocean **tsunami** pounded the coast-

Toxic Waste Disasters 23

lines of eleven countries, from Thailand to Africa. One African country that suffered serious damage in its northern region was Somalia. About 300 people were killed and thousands of homes were destroyed. And even though no one knew it at the time, the tsunami had also caused another type of problem. As powerful waves pounded the shoreline, they stirred up old rusting barrels that contained tons of toxic waste. The barrels had been illegally dumped in the ocean during the 1980s and 1990s, and the force of the waves broke them open.

About two months after the tsunami struck, huge amounts of toxic waste began washing ashore

Broken concrete and other rubble covered Somalia's coast after the 2004 tsunami. Barrels of toxic waste also washed up on Somalia's beaches.

in Somalia. The beaches became littered with radioactive waste and heavy metals such as cadmium, mercury, and lead. There were also numerous chemicals, infectious medical waste, and other toxic substances. Health-care workers found that people in nearby villages had begun suffering many health problems. For instance, there were many respiratory infections, unusual skin disorders, internal bleeding, and bleeding mouth sores. Officials say the health problems could grow much worse as time passes. It will likely be many years before the full effects of the toxic waste are known.

In the aftermath of Hurricane Katrina, much of New Orleans was submerged in a toxic soup.

"Toxic Soup"

Eight months after the tsunami devastated Southeast Asia, a different kind of natural disaster struck the United States—and it, too, created a toxic waste crisis. On August 29, 2005, a vicious hurricane named Katrina slammed into the southeastern coast from the Gulf of Mexico. One of the areas that was hardest hit was New Orleans, a bowl-shaped city that lies below sea level. Katrina pounded the city with driving rains and winds of over 100 miles (161k) per hour . Before long, protective barriers known as **levees** collapsed. Water started pouring into the city from Lake Pontchartrain, a massive lake to the north. Within 24 hours, about 80 percent of New Orleans was underwater. Hundreds of people drowned in the flood, including many who were trapped inside their own homes.

Over the following days, the water covering the streets of New Orleans became polluted with deadly toxins. Explosions at a chemical storage facility and railroad yard spewed toxic fumes into the air and chemicals into the water. Tons of garbage spilled out of flooded dumpsters, and broken sewer pipes gushed raw sewage. Thousands of gallons of oil were released from two oil spills, while fuel, battery acid, and other toxic substances leaked from gas stations and overturned vehicles. In the midst of it all, unknown numbers of dead, decomposing humans and animals floated in the water.

As the days went by, the water in New Orleans continued to grow more rancid, and people referred to it as "toxic soup". An article on CNN.com described what it was like: "The toxic nature of the water is evident from the smell of garbage, human waste and rotting corpses, and the slick sheen of oil, gasoline and other chemicals on the surface."[6] Health officials warned that there was an enormous health risk for anyone who had come in contact with the polluted water—and that included thousands of people. Mayor Ray Nagin ordered all residents to evacuate until the water was gone and the city could be cleaned. But as of November 2005, 22 million tons (20 million metric tons) of debris— enough to fill the Superdome 43 times—remained in the city, and millions of tons of hazardous waste still needed to be disposed of. No one knows for sure when New Orleans will be declared completely safe again. An even bigger unknown is how serious the long-term effects will be on human health.

Brazilian Catastrophe

An equally devastating toxic waste crisis occurred in Brazil in 2003. This one, however, was not caused by a natural disaster. In late March, the wall of a waste reservoir at a paper manufacturing plant north of Rio de Janeiro collapsed. More than 300 million gallons (1.2 billion liters) of **caustic** soda, chlorine, and other toxic chemicals spilled out of the reservoir and into one of Brazil's major

Toxic Waste Disasters 27

Poisoned by a toxic spill at a paper manufacturing plant in Brazil, dead fish pile up along the shore of a popular beach.

rivers, the Paraíba do Sul. The poisonous waste soon spread into another Brazilian river, the Pomba, before the spill could be stopped.

Like many other paper manufacturers, the Brazilian mill had used the chemicals for bleaching wood pulp. The resulting waste material was strong enough to burn the skin off any living thing. The toxic chemicals coated the rivers with a thick layer of smelly foam that spread rapidly and killed everything in its path. Soon the water

Toxic Waste

Cleanup begins in Romania in 2001 after a chemical factory spill. One year earlier, toxic waste from a gold mine poisoned Romania's rivers and streams.

was littered with the bodies of thousands of dead fish. In addition, the toxic foam killed hundreds of cattle and any other animals that drank the water, as well as all vegetation in the water. Within days, the rivers were completely empty of life. A half-million people had no access to clean water for drinking or bathing for more than a month. As for the long-term damage, Brazilian officials say it could take as long as fifteen years before the full effects of the environmental disaster are known.

Deadly Mining Spill

The east European country of Romania experienced a toxic waste spill that was also caused by a collapsed reservoir wall. The accident happened on January 30, 2000, at the Baia Mare gold mine near the border town of Oradea. More than 20,000 tons (18,000 metric tons) of cyanide, heavy metals, and other toxic substances poured out of the mine's waste reservoir and into Romania's Somes River.

The damage did not stop in Romania, however. The toxic waste continued traveling for hundreds of miles, poisoning rivers and streams all along the way. Major rivers in Hungary and Yugoslavia were polluted by the spill, as was the Black Sea, where the rivers emptied. The mayor of one Yugoslavian town described the damage there: "This is not just an environmental crisis. This is the poisoning of a river. Eighty percent of everything in the river will die. The wave of poison will pass within 36 hours, but no one here knows how to cope with the catastrophe."[7] Because the damage to the environment and wildlife was so great, people often call the Romanian chemical spill one of the worst ecological disasters in Europe's history.

Catastrophic toxic waste accidents such as these are rare. But when they do happen, they can lead to enormous damage. Whether they are caused by large chemical spills or natural forces, their effects can linger in the environment for many, many years—and often, there is no possible way to predict just how serious the long-term damage will be.

Chapter Four

Is There an Answer?

Although toxic waste remains a serious environmental problem, scientists have made progress in studying and understanding it. New technologies have been developed to help companies reduce toxic waste and make their emissions cleaner. Recycling operations help reduce the amount of e-waste that ends up in landfills. Communities regularly sponsor programs for people to drop off used oil, paint, and other toxic waste so that it can be disposed of properly. Yet in spite of the progress that has been made, there are still many challenges ahead. One of the biggest challenges is cleaning up toxic waste contamination that happened long ago.

Messes from the Past

In the United States, the Environmental Protection Agency (EPA) oversees these cleanup operations. The agency is responsible for administering the Comprehensive Environmental Response, Compensation, and Liability Act, more commonly known as Superfund. The Superfund program was put into law in 1980, and was designed to ensure that contaminated sites are found, investigated, and cleaned up. That is an enormous undertaking because there are more than 1,200 of these sites all over the country. It costs hundreds of millions of dollars to handle such a large cleanup effort. So getting the work done is a time-consuming, tedious, and extremely expensive process.

Workers in protective clothing carefully handle hazardous waste.

One of the highest-priority Superfund sites is in Tar Creek, Oklahoma. Zinc and lead mining companies operated there from the early 1900s until the 1960s. When the operations closed, they left huge piles of waste materials behind, some of them more than 200 feet (61m) high. The companies also left large pools of poison-laced water. The toxic waste has polluted the soil in Tar Creek, as well as the groundwater. As a result, more than 30 percent of the children living there have abnormally high levels of lead in their blood. The U.S. government has already spent millions of dollars in cleanup costs in Tar Creek. However, the area remains highly contaminated, so there is still much work to be done.

Safer Disposal

Tar Creek and other Superfund sites are problems left over from the past. Years ago, companies were often careless with their disposal of toxic waste. That was largely because laws designed to protect the environment did not exist. Today there are a number of environmental protection laws. One area that is strictly regulated is hazardous waste disposal. Only certain landfills are allowed to accept toxic waste, and they must meet strict standards to protect the environment and public health. These landfills have heavy-duty structures and multiple reinforced liners to protect against leakage. The people who run the operations closely

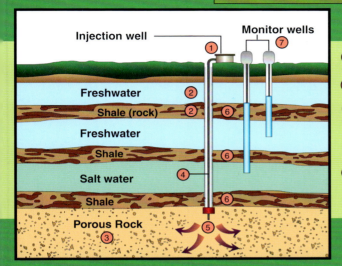

Deep Well Injection
ONE METHOD OF TOXIC WASTE DISPOSAL

① An injection well is drilled many thousands of feet into the ground, ② through layers of water and shale rock, until ③ it reaches porous rock. ④ Toxic waste is pumped into the well. ⑤ The waste disperses into the porous rock, where it replaces natural gases. ⑥ The weight of the rocks above keep the waste in place. ⑦ Monitor wells make sure that toxic waste has not seeped upward and polluted the freshwater.

monitor the toxic waste that is disposed of there, and the companies that drop it off. The Hazardous Waste Resource Center describes this type of facility: "Landfills that can accept hazardous waste are as different from the old county dumps are as a tricycle is to a new Cadillac."[8]

When landfills are not suitable for certain kinds of liquid toxic waste, a type of disposal called **deep well injection** may be used. With this method, wells are drilled thousands of feet into the ground until they hit a layer of porous rock. The toxic waste is then pumped into the cavity, where it replaces natural fluids or gases. The waste remains trapped in place by immense pressure from the rocks above it. This method is used in Texas more than in any other state. In 2001 industries in Texas injected over 77 million pounds (35 million kg) of toxic chemicals and compounds deep into the earth.

Model Industries

While improved disposal methods have helped address the problems of toxic waste, most scientists agree that the ideal solution is developing technologies that keep the waste from being produced in the first place. Reducing toxic waste is not an easy task, though. It can be very expensive for businesses to develop new technology. It

Lockheed Martin builds military aircraft such as the F-22 Raptor (pictured). The company cut toxic waste by switching cleaning solutions.

means they must make major changes in the way they run their operations. However, many companies are so committed to protecting the environment that they are willing to make those changes.

One example is an industrial gas producer called Air Products and Chemicals, which is located in Allentown, Pennsylvania. The company invested in technology that helped cut hazardous-waste generation by more than half. In some of its plants, Air Products and Chemicals has reduced its toxic air emissions by 60 percent. One of its chemical plants provides a neighboring company with waste that is used for fuel. This reduces the neighboring facility's requirements for traditional energy sources.

Another corporation that works hard to protect the environment and reduce toxic waste is automaker Toyota. Since 2000 the company has reduced its hazardous waste by 40 percent. Toyota has also invested in technology that helps employees monitor waste-related data. The data is used to find ways of making even more reductions in the waste that is produced at the Toyota manufacturing plants.

Lockheed Martin, a military aircraft manufacturer, reduced its toxic waste by switching to a new kind of cleaning product. Its equipment maintenance shops were using many different hazardous solvents to clean parts. More than twenty of the shops were using nearly 1,000 gallons (3,800 liters) of chemical solvents every year. Lockheed Martin wanted to find a way to achieve the same cleanliness without using products that were hazardous to employees and the environment. After doing some research, the company discovered a new type of solvent. It is water-based, rather than made with toxic chemicals. The new cleaner is unique because it contains enzymes that eat grease, oil, and other impurities. This is a process known as **bioremediation**, and its potential goes far beyond cleaning metal. Scientists say it could play a major role in helping to solve the world's hazardous-waste problems.

Good Bacteria

Bioremediation uses microscopic organisms such as bacteria to speed up the rate at which substances biodegrade, or naturally break down. One especially promising feature is that it works very well on waste products that are buried in sediments. So the technique may prove to be very effective on waste in landfills.

One of the first times bioremediation was used was in the early 1990s, following the disastrous *Exxon Valdez* oil spill in Alaska. Scientists applied fertilizer to certain areas of the Alaskan shoreline. Over time, the bacteria began to eat the oil. That was good news for scientists because it meant bioremediation was helping nature do its job faster.

Is There an Answer? 37

A Drifting Oil Slick

Exxon Valdez runs aground March 24, 1989

- Anchorage
- Valdez
- Whittier
- Kenai Peninsula
- Seward
- Cook Inlet
- Homer
- Alaska Peninsula
- Gulf of Alaska
- Kodiak Island

DAY 4 March 27 — 37 miles
DAY 7 March 30 — 90 miles
DAY 11 April 3 — 140 miles
DAY 14 April 7 — 150 miles
DAY 19 April 14 — 250 miles
DAY 40 May 2 — 350 miles
DAY 56 May 18 — 470 miles

The *Exxon Valdez* oil spill was one of the first times bioremediation was used to aid in cleanup.

Source: Idaho State University, Department of Geosciences, http://wapi.isu.edu.

The *Exxon Valdez* makes its way to shore following its disastrous oil spill.

In the spring of 2005, scientists from the University of South Carolina performed some experiments of their own with bioremediation. They discovered a type of powerful bacteria that could eat even the most toxic pollutants, such as PCBs and many hazardous chemicals. But they also discovered something else about the bacteria—not only could it eat toxic waste, it could also generate electricity! One of the researchers, Charles Milliken, said enough power could be generated by bioremediation to operate small electronic devices. He explains: "As long as the bacteria are fed fuel they are able to produce electricity 24 hours a day."[9]

The Future

Whether toxic waste is left over from the distant past or the result of modern industry, dealing with it is a challenge. Scientists have made progress by developing better disposal methods and new technologies. Major corporations have shown their commitment to a cleaner environment by investing in those technologies. Environmental laws help prevent the problems of the past from occurring again. But the question is: Will tomorrow bring a world that is free from toxic waste? No one can say for sure. Hopefully, with all that has been learned along the way, people everywhere will agree that no cost is too great to protect the planet from toxic waste contamination.

Notes

Chapter 1: Hazardous Ooze

1. Greenpeace, "The E-Waste Problem," March 8, 2005. www.greenpeace.org/international/campaigns/toxics/electronics/the-e-waste-problem.

Chapter 2: How Serious Is the Problem?

2. Kentucky Environmental Quality Commission, "Teacher's Guide to Kentucky's Environment," updated December 29, 2004. www.eqc.ky.gov/pubs/tg.
3. Quoted in BBC News, "Coca-Cola's 'Toxic' India Fertiliser," July 25, 2003. www.indiaresource.org/news/2003/4703.html.
4. Quoted in Dan Kennedy, "The *New Yorker* Ignores Leukemia in Woburn," *Boston Phoenix*, February 12, 1999. http://home.earthlink.net/~dkennedy56/phoenix_990212woburn.html.
5. Jane Houlihan, Timothy Kropp, Richard Wiles, Sean Gray, and Chris Campbell, "Body Burden: The Pollution in Newborns," *Environment Working Group Report,* July 14, 2005. www.ewg.org/reports_content/bodyburden2/pdf/bodyburden2_final-r2.pdf.

Chapter 3: Toxic Waste Disasters

6. CNN.com, "New Orleans Will Force Evacuations," September 7, 2005. www.cnn.com/2005/US/09/06/katrina.impact.
7. Quoted in Lisa Gurevitch, "Cyanide Spill Pours Across Border into Yugoslavia," CNN.com, February 13, 2000. http://edition.cnn.com/2000/WORLD/europe/02/13/yugo.cyanide.

Chapter 4: Is There an Answer?

8. Hazardous Waste Resource Center Web site, "Engineered Landfills." www.etc.org/technologicalandenvironmentalissues/treatmenttechnologies/landfills/index.cfmP.
9. Quoted in *LiveScience,* "Double Bonus: Bacteria Eat Pollution, Generate Electricity," June 7, 2005. www.livescience.com/technology/050607_bacteria_electricity.html.

Glossary

bioremediation: The use of microorganisms such as bacteria to remove toxins from the environment.

caustic: Strong enough to burn or destroy animal flesh or tissue.

deep well injection: A method of disposal that injects toxic wastes deep into the ground.

dioxins: Deadly chemical by-products created during certain manufacturing processes or incineration.

e-waste: A term used to describe discarded computers, cell phones, and other electronic equipment.

food chain: A community of plants and animals in which each member is eaten in turn by another member.

groundwater: Water supplies that are found deep beneath the Earth's surface.

heap leaching: A process used by mining companies to extract gold or other minerals by spraying piles of ore with chemicals such as cyanide.

heavy metals: Lead, cadmium, mercury, arsenic, and other substances that can be highly toxic.

hydrocarbons: Chemical compounds found in petroleum products.

levees: Protective barriers built along the banks of a river or lake to protect against flooding.
radioactive waste: Waste that is generated during the production of some type of nuclear energy.
sludge: Gooey, thick deposits or sediments that are sometimes toxic to the environment.
toxic waste: Waste that is poisonous to humans, wildlife, and the environment.
tsunami: A series of massive sea waves caused by violent disturbances in the Earth.

For Further Exploration

Books

Nichol Bryan, *Love Canal: Pollution Crisis.* Milwaukee: World Almanac Library, 2004. The story of one of the most famous toxic waste crises of all time. The author discusses what happened, and covers such issues as health risks when an area is contaminated, how tests are performed, and cleanup methods.

August Greeley, *Toxic Waste: Chemical Spills in Our World.* New York: PowerKids Press, 2003. This book discusses the many ways that chemical products are used today, and explores the impact of spills on the environment, wildlife, and humans.

Rosie Harlow, *Garbage and Recycling.* New York: Kingfisher, 2001. Explains how many different kinds of waste can be recycled, and includes tips for how young people can have a positive impact on the planet by recycling.

Periodical

Social Issues Resources Series (SIRS) staff, "Toxic Wastes: Difficult to Get Rid Of," *SIRS Digests*, Fall 1996.

Web Sites

EcoKids (www.ecokids.ca). This site by Earth Day Canada teaches young people about the environment through interactive educational games and activities. Kids are encouraged to form their own opinions, get involved, and learn to understand the impact their own actions have on the environment. Includes a section devoted to toxic waste.

Environmental Protection Agency Environmental Kids Club (www.epa.gov/kids). Features animated stories, games, and many different fun activities for kids interested in learning about the environment. The site includes a whole section devoted to garbage and recycling.

National Institute of Environmental Health Sciences Kids' Pages (www.niehs.nih.gov/kids). This informative site teaches young people about toxic waste and other issues with a variety of games and activities. Also includes a collection of links to other good environmental sites.

Planetpals (www.planetpals.com). A wonderful collection of environmental information especially for kids. One feature of the site is the *Planetpals Earthzine,* a colorful online magazine designed to make learning fun.

Rotten Truth About Garbage (www.astc.org/exhibitions/rotten/rthome.htm). This site by the Association of Science-Technology Centers is designed to educate young people about waste, disposal methods, and recycling.

Index

air pollution, 6, 35
Air Products and
  Chemicals, 35
Alaska, 36
Allentown, Pennsylvania, 35
aluminum, 13
Alzheimer's disease, 13
arsenic, 7

babies, 18–21
battery manufacturing, 7
bioremediation, 36, 38
Black Sea, 29
Brazil, 26–28

cadmium, 7
  in electronic equipment, 9
  health problems from, 13
California Public Interest
  Research Group
  (CALPIRG), 16
cellular phones, 9
children, 18–21, 32
chlorine, 6
chromium, 7
cleaning products, 35–36
CNN.com, 26
Comprehensive
  Environmental Response,
  Compensation, and
  Liability Act (1980), 31–32
computers, 9
Condon, Suzanne, 20
cyanide, 8–9, 13

deep well injection, 33–34
dioxins, 6
  in fertilizers, 16–18
  health problems from, 13
disposal, safer, 32–34

electronic equipment, 9
energy sources, 35, 38
Environmental Protection
  Agency (EPA), 31–32
Environmental Working
  Group (EWG), 21
Everglades National Park
  (Florida), 7
e-waste, 9–10
Exxon Valdez, 36

farms, 12, 16–18
fertilizers, 12, 16–18
fetuses, 21
fish, 15–16
Florida, 7

45

food chain, 15–16

Greenland, 16
Greenpeace International, 9
groundwater, pollution of, 9, 10, 32

Hazardous Waste Resource Center, 33
health-care facilities, 11–12
health effects, 13, 18–21, 24
heap leaching, 8–9
heavy metals
  in e-waste, 9
  in fertilizers, 17–19
  health problems from, 13, 18–21
  from industry, 7
  see also specific metals
Henry, John, 17
household products, 11
Hungary, 29
hydrocarbons, 11

industry
  reduction of toxic waste by, 35–36
  toxic waste from, 5–7, 26–28

Katrina (hurricane), 25–26
Kentucky, 14–15
Kerala (India), 17

landfills
  e-waste in, 10
  leakage of waste from, 14–15
  standards for, 32–33
lead
  from battery manufacturing, 7
  blood levels in children, 32
  in electronic equipment, 9
  health problems from, 13
  from mining, 7
leukemia, 19–21
Lockheed Martin, 35–36

marine mammals, 15–16
Massachusetts Department of Public Health (DPH), 20
Maxey Flats, Kentucky, 15
mercury
  in air, 7
  in electronic equipment, 9
  in farming, 16–18
  in food chain, 15–16
  health problems from, 13
  in mining, 7, 9
metals. *See* heavy metals; *and specific metals*
Milliken, Charles, 38
mining, 7–9, 28–29, 32
motor oil, 11

Nagin, Ray, 26
National Academy of Sciences, 11
New Orleans, Louisiana, 25–26
nuclear power plants, 11

oil, 11, 36
oil sludge, 11
open-pit mining, 7–8

Index

paper mills, 6, 26–28
Paraíba do Sul River, 27
Parkinson's disease, 13
power plants, 7, 35, 38

radioactive waste, 11, 15
recycling, 30
reduction technologies, 34–36, 38
Romania, 28–29

sludge, 7, 11
solutions
  cleanup, 31–32
  disposal, 32–34
  recycling, 30
  reduction technologies, 34–36, 38
Somalia, 23–25
Somes River, 29
sources
  disposal of by individuals, 9–11
  farms, 12
  health-care facilities, 11–12
  illegal dumping, 23
  industry, 5–7
  mining, 7–9, 28–29, 32
  after natural disasters, 23–25
  nuclear power plants, 11
Superfund, 31–32

Tar Creek, Oklahoma, 32
televisions, 9
Texas, 33–34

Toyota, 35
tsunami of 2004, 23

United Nations Environment Programme (UNEP), 7, 16
United States
  amount of waste created in, 4
  cleanup in, 31–32
  landfill standards in, 32–34
University of South Carolina, 38

Valley of the Drums (Kentucky), 14–15

waste products, amount of, 4, 9
wastewater, 6
water pollution
  from e-waste, 10
  health effects of, 19–21
  from household products, 11
  from industry, 6, 26–28
  from mining, 8, 9, 28–29, 32
  from waste sludge, 17–18
Woburn, Massachusetts, 19–21

Yugoslavia, 29

Picture Credits

Cover photo: © age fotostock/SuperStock
Maury Aaseng, 14, 17, 19, 33
AP/Wide World Photos, 28
© Francesco Broli/World Food Programme/Pool/Reuters/CORBIS, 23
© Bettmann/CORBIS, 18
Centers for Disease Control, 31
© Roy Corral/CORBIS, 37 (photo)
© 2004 Lockheed Martin/Getty ImagesNews/Getty Images, 34
© Alberto Moreno/Reuters/CORBIS, 10
National Geographic Channel/EPA/Landov, 20 (inset)
PhotoDisc, 6 (both photos), 12 (both photos), 20 (large photo)
© Reuters/CORBIS, 27
© David Sailors/CORBIS, 5
© Paul A. Souders/CORBIS, 8
© Rick Wilking/Reuters/CORBIS, 24

About the Author

Peggy J. Parks holds a bachelor of science degree from Aquinas College in Grand Rapids, Michigan, where she graduated magna cum laude. An avid environmentalist and nature lover, Parks has written more than 40 books for Thomson Gale's KidHaven Press, Blackbirch Press, and Lucent Books imprints. She lives in Muskegon, Michigan, a town she says inspires her writing because of its location on the shores of Lake Michigan.

Toxic waste
31290093562916 BH